CONTENTS

This book looks at the three planets that are nearest the Earth in the Solar System, Mercury, Venus and Mars. Mercury and Venus both lie closer to the Sun than the Earth, while Mars lies further away. All three planets are made up mainly of rock like the Earth, and astronomers call them the terrestrial, or Earth-like planets. But these planets are unlike the Earth in most other ways. Mercury is a searing hot world, baked by the nearby Sun. Venus gets equally hot because its heavy atmosphere traps the Sun's heat like a gigantic greenhouse.

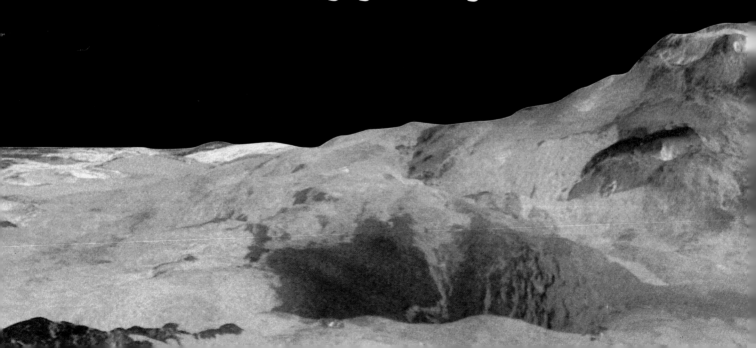

NEAR PLANETS

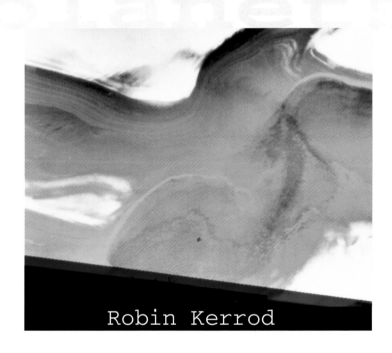

Robin Kerrod

Belitha Press

First published in Great Britain in 2000 by

Belitha Press
A member of Chrysalis Books plc
64 Brewery Road, London N7 9NT

Paperback edition first published in 2003
Copyright © Belitha Press Limited 2000
Text by Robin Kerrod

Editor: Veronica Ross
Designers: Caroline Grimshaw, Jamie Asher
Illustrator: David Atkinson
Consultant: Douglas Millard
Picture researcher: Diana Morris

ISBN 1 84138 061 X (hb)
ISBN 1 84138 755 X (pb)

British Library Cataloguing in Publication Data for this book
is available from the British Library.

Printed in Hong Kong

10 9 8 7 6 5 4 3 2 1 (hb)
10 9 8 7 6 5 4 3 2 1 (pb)

Picture credits

Robin Kerrod/Spacecharts: 10, 11.
NASA/Spacecharts: 1, 3, 4, 5, 9, 13, 15, 16, 17, 18, 19, 21tr, 22, 24-5,
25, 26, 27, 28, 29, 31tr, 32 background, 34, 35, 36, 37,
38, 39, 40, 41, 42-3.
National Museum of Archaeology, Naples/Erich Lessing/AKG
London: 20tr.
Michael Teller/AKG London: 30tr.
Tempio Malatestiano, Rimini/Bridgeman Art Library: 12tr.

Some of the more unfamiliar words used in this book are
explained in the glossary on pages 46 and 47.

But Mars is much colder than the Earth, with freezing temperatures for most of the time. Each of the near planets has a different kind of surface from the Earth and from each other. Mercury has an old, heavily cratered surface. The surface of Venus, on the other hand, is relatively young. There are only a few craters but plenty of volcanoes. Mars has craters and volcanoes, all of them ancient. The planet is a rusty red colour, and is known as the Red planet.

NEIGHBOURS

Earth and its planetary neighbours lie near the Solar System's centre.

The Earth and eight other planets form the main part of the Solar System, the family of bodies that travels through space with the Sun. The planets are spread out over distances so great that they are almost impossible to imagine. Even Venus, the planet that comes closest to the Earth, never comes nearer than about 42 million kilometres. Yet we can consider it a neighbour among the planets.

Mars and Mercury are further away than Venus. But they are both close enough to be called neighbours when we see how far away the other planets are. The other planets lie hundreds of millions and even thousands of millions of kilometres away. If we set off in the space shuttle to visit Venus, it would take us only about two months. But it would take us 25 years to reach Pluto, the most distant planet.

▽ **Earth and its three neighbours lie much closer together than the other planets. But all the planets travel around the Sun in the same direction.**

Earth

Mars

Mercury

Sun

Venus

orbit of Jupiter

Jupiter

orbit of Saturn

Saturn

Uranus

orbit of Uranus

Pluto

orbit of Neptune

Neptune

orbit of Pluto

rocky lumps
stick together

planet
forms

planet's
surface melts

gas
blown away

△ **Earth and its neighbours formed when lumps of matter came together under gravity.**

How the planets formed

The Solar System was born about 4600 million years ago, when a great cloud of gas and dust in space began to collapse. Inside the cloud, particles of gas and dust were pulled together by gravity to form a hot ball, with a disc around it. The hot ball became the Sun, and matter in the disc came together to become the planets.

The four planets closest to the Sun – Mercury, Venus, Earth and Mars – started to form when bits of rocky matter collided with one another and stuck together to form larger and larger lumps. In time the lumps grew in size to become the planets.

The gas that was in the inner part of the disc was soon blown away by the force of the rays and particles given off by the new-born Sun. It was blown into the colder, outer reaches of the Solar System, where it became part of the giant outer planets.

The four inner planets went on to develop in quite different ways to produce the four quite different bodies we find today.

Close companion

Venus is the closest planet to the Earth, but it isn't our closest neighbour in space. That is the Moon, which comes as close to Earth as 360 000 kilometres. Some of the mini-planets we call the asteroids also sometimes pass within a few million kilometres of the Earth. A few come even closer. We call them the Earth-grazers.

PLANETS LIKE EARTH

In make-up, the near planets have much in common with the Earth.

Mercury, Venus and Mars are known as the terrestrial, or Earth-like planets. This is not because they look like the Earth, but because they are similar to the Earth in make-up.

The Earth is made up of rock and metal, mainly iron. Geologists – the scientists who study the Earth – tell us that the Earth is made up of a number of different layers. They found this out by studying the way earthquake waves travel through the Earth.

On top is the crust, a thin layer of hard rock. Underneath there is a much thicker layer of heavier rock, called the mantle. Underneath that, in the centre of the Earth, is the core, which is made up mainly of iron. The outer part of the core is liquid.

Inside the planets

Astronomers cannot find out directly what the terrestrial planets are like inside. But they think that the planets must be similar to the Earth because they formed at the same time and in the same way. They have a crust and a mantle made up of rock, and an iron core. The size of these layers varies from planet to planet.

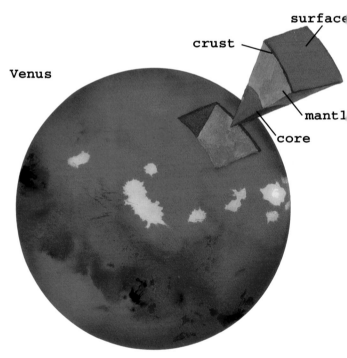

Venus

surface
crust
mantle
core

Mercury

surface
crust
mantle
core

△ Venus probably has a mantle and core about the same size as those of the Earth.

◁ Mercury has a thin mantle but a thick core.

Venus is nearly the same size as the Earth, so its structure is probably very similar. But its core is probably slightly smaller and completely solid. Mercury, on the other hand, seems to have a very big core for its size, and only a thin layer of mantle. Mars has a smaller core but a thicker mantle.

Atmospheres

Another major difference between these planets is their atmospheres, or the layer of gases that surround them. Mercury has no atmosphere, and Mars has only a trace of one. But Venus has a thick atmosphere – much thicker than the one on Earth.

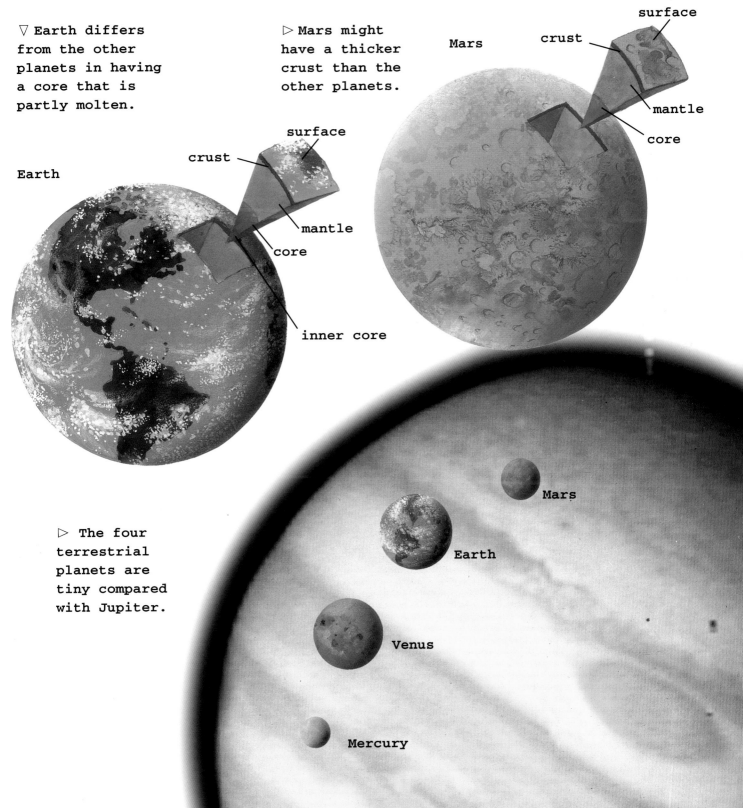

▽ **Earth differs from the other planets in having a core that is partly molten.**

▷ **Mars might have a thicker crust than the other planets.**

Mars

surface
crust
mantle
core

surface
crust
mantle
core
inner core

Earth

▷ **The four terrestrial planets are tiny compared with Jupiter.**

Mars

Earth

Venus

Mercury

WANDERING STARS

All three near planets can be seen shining like stars in the night sky.

△ **Venus shines in the dawn sky as the morning star.**

▽ **A planet can be a morning star (left) or an evening star (right) depending on which side of the Sun it is.**

On many evenings of the year, a bright star appears in the twilight of the western sky. Only later, when the sky darkens, do the other stars come out. But this bright evening star is not a star at all. It is the planet Venus.

In a similar way, the two other near planets, Mercury and Mars, can at times be seen shining like bright stars in the sky. But, even though they look like stars, the three planets are quite different from real stars.

Real stars remain fixed in their constellations, or the star patterns we see in the sky. The planets change their positions among the constellations all the time. The ancient astronomers saw this, which is why they called them planets, a Greek word meaning 'wanderers'.

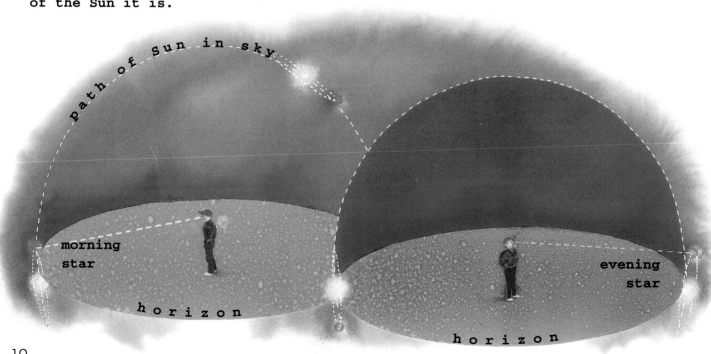

Path of Sun in sky

morning star

horizon

evening star

horizon

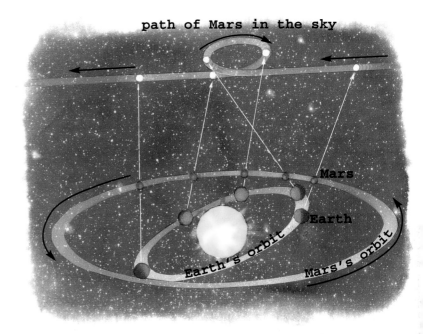

△ **Seen from the Earth, Mars sometimes appears to travel backwards through the heavens.**

Spot the difference

You can see another basic difference between a planet and a star by looking at Venus in binoculars or a telescope. The stars show up only as little pinpricks of light, but Venus shows up as a distinct circle, or disc.

Venus (and the other planets) look bigger than the stars only because they are very much closer to us – only a few tens of millions of kilometres. The stars look so tiny because they are millions of times further away. In fact they are hundreds of times bigger than the planets.

There is also another important difference between the planets and real stars. Real stars are searing hot bodies like the Sun, which give off light of their own. Planets give off no light of their own. They shine because they reflect the light of the Sun.

Planet-watching

The three near planets all look quite different in the sky. Venus is easiest to spot because it is so bright. It is most familiar as the evening star, which appears in the west just after sunset. But at some times of the year, it can be seen as a morning star in the east just before sunrise. Mercury too can be a morning or an evening star. But it is more difficult to see.

Mars, on the other hand, is usually found in the dark skies of night. It does not shine brightly all the time. Only when it comes close to Earth does it shine brighter than the stars. Then we can easily recognize it by its reddish colour.

◁ **Mercury appears in the sky just after sunset as an evening star. It is always seen close to the horizon.**

MERCURY: THE SPEEDY PLANET

Being closest to the Sun, Mercury travels fastest in its orbit.

△ **The symbol for Mercury.**

People began observing Mercury thousands of years ago. The Greeks called the planet Hermes, after the swift messenger of the gods in Greek mythology. The Romans called him Mercury.

Planet Mercury is difficult to see because it is always close to the Sun when we view it from the Earth. This means that at sunrise or sunset, it is never far above the horizon. And it is never seen in a dark sky. But under ideal conditions, it outshines all the stars, even the brightest one, Sirius.

△ **A carved relief showing Mercury.**

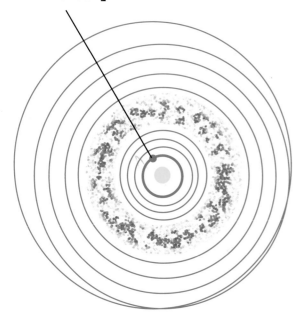

△ **Mercury is the planet that is closest to the Sun.**

Travelling at speed

Mercury travels faster in its path, or orbit, around the Sun than any other planet. It takes just under 88 days to circle the Sun, at an average speed of nearly 173 000 kilometres an hour.

Most planets travel around the Sun in a nearly circular orbit. But Mercury travels in a more oval, or elliptical orbit. This means that it is much closer to the Sun at some times than at others. This causes its speed to change. It gets faster when it nears the Sun and slower when it goes further away. If you lived on Mercury, you would find this change in speed would cause some odd effects. At certain times the Sun would appear to move backwards in the sky. At others the Sun would rise, set, then rise again on the same day.

Mercury

Earth

Moon

△ With a diameter of 4878 km, Mercury is much smaller than the Earth and only a little bigger than the Moon.

Mercury's long days

All planets spin round in space like a top as they travel in orbit around the Sun. The Earth spins round once a day. The other planets spin round in different periods of time.

Mercury spins round very slowly. It takes nearly 59 days to spin round once. This slow spin gives Mercury a very long 'day', or time between sunrise and sunset. Its 'day' lasts 88 Earth-days. Its 'night' is also 88 Earth-days long.

△ Ancient craters cover most of Mercury, making it look much like the Moon. This image is a mosaic of hundreds of separate smaller pictures taken by the space probe *Mariner 10*.

Terrific temperatures

The long days and nights make Mercury at the same time scorching hot and freezing cold. During the day, the Sun blazes down relentlessly, baking the surface to temperatures up to about 450 degrees Celsius (°C). This is high enough to melt the metals lead, tin and zinc.

When the Sun goes down and night falls, temperatures plunge rapidly. By the end of the long night, the dark surface of Mercury is a deep-frozen −180°C. On Earth, this would be nearly cold enough to turn the air that we breathe into liquid.

MAPPING MERCURY

Mercury looks just like the Moon.

When you look at Mercury in a telescope, you see a small white disc. In a large telescope, you might be able to see some vague markings on its surface. But even the most powerful telescope will not be able to show any more details because the planet is too small and too far away.

Astronomers had no idea what Mercury was really like until an American space probe flew past it in March 1974. The probe was *Mariner 10*. It photographed the surface with television cameras and flashed the pictures back to Earth.

▷ **This map is based on the pictures sent back by *Mariner 10*. Planitia means a plain. Some large craters are named after famous writers, artists and musicians.**

Caloris Planitia

Tolstoy

Mariner's mission

Pictures showed that Mercury's surface is peppered with thousands of craters. It looks very much like the surface of the Moon.

After passing close to Mercury, *Mariner 10* went into orbit around the Sun. Six months later, its orbit took it back past Mercury again, where it took more pictures. Six months later, it returned for a third and final look at the Moon-like planet. On its three flypasts, *Mariner 10* took more than 2500 pictures that covered nearly half the planet's surface.

magnetometers

antenna

TV cameras

solar panel

sunshade

thrusters

steerable dish antenna

◁ **The space probe *Mariner 10*. It took pictures of Mercury with a pair of television cameras.**

14

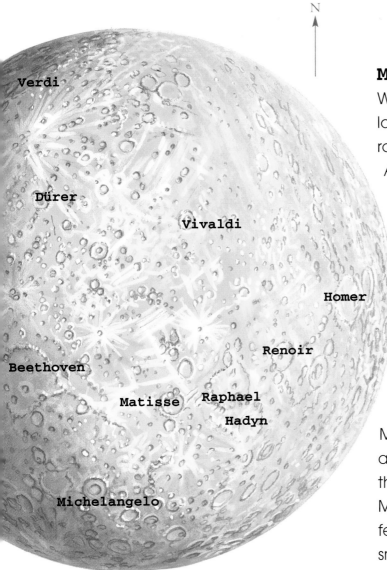

Verdi

Dürer

Vivaldi

Homer

Renoir

Beethoven

Matisse Raphael

Hadyn

Michelangelo

N

Mimicking the Moon

We should not be surprised that Mercury looks so much like the Moon. They are both rocky bodies and are about the same size. Also, they have no atmosphere, or layer of gases around them.

Both bodies were heavily bombarded by rocks from outer space when they were very young, which dug out the pits, or craters, that cover their surfaces. Their surfaces are very ancient and have changed little in billions of years.

But, there is one major difference between Mercury and the Moon. The Moon has many large dusty plains, regions astronomers call seas, or *maria*. These are the dark areas we see when we look at the Moon in the night sky. Mercury has only a few small plains and they are not as smooth as the ones on the Moon.

▽ **The cratered surface of Mercury.**

△ **The cratered surface of the Moon.**

THE CRATERED SURFACE

Craters large and small cover nearly the whole surface of Mercury.

Mercury is the most heavily cratered of all the terrestrial planets. Its neigbours in space, Venus and the Earth, have only a few craters on their surface. Why is Mercury so different?

Long ago Venus and the Earth were covered in craters like Mercury. All three planets were heavily bombarded by lumps of rocks, or meteorites, soon after they formed (see page 7). But over the years processes have been at work on Venus and the Earth that have covered up or got rid of most craters.

△ One of the flatter plains regions of Mercury. The big crater on the left is about 30 km across.

▽ The southern hemisphere of Mercury, pictured by *Mariner 10*. The largest of the craters near the bottom is about 170 km across.

Crater shapes and sizes

Meteorites rained down on Mercury long ago and dug out different-sized craters. The smallest ones are bowl-shaped and measure up to a few kilometres across. Larger ones have walls that slope down in terraces to a flat floor. Mountain peaks rise up in the middle of the floor. This kind of terraced crater is also found on the Moon. Very large rocks dug out even bigger craters, in which the floor rucked up to form rings of mountain peaks.

△ **Many large craters on Mercury have terraced walls and mountain peaks in the middle.**

Changing the landscape

The main processes have been volcanism and erosion. Volcanism is the eruption of volcanoes, which pour molten rock, or lava, over the landscape. Erosion is the gradual wearing away of the surface, by the weather, for example. And on Earth another process that has changed and is still changing the landscape is continental drift. This is the movement of sections, or plates, of the crust.

On Mercury, volcanoes have not erupted for billions of years. Erosion by the weather has not happened because the planet has no atmosphere. And it has had no moving plates. While the other planets have changed, Mercury remains as it was four billion years ago – covered with craters.

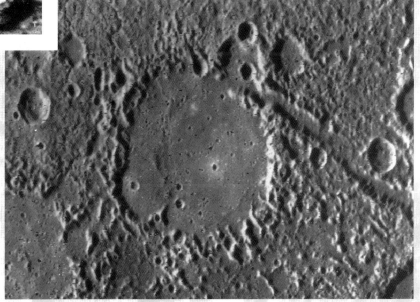

△ **One of Mercury's valleys. It is more than 100 km long.**

Mercury's plains

Mercury does not have the large plains, or 'seas', that the Moon has. But it does have a few quite flat areas. They are known as planitia. (They are named after certain gods, for example, Odin Planitia, after the Norse god Odin.) As well as the planitia, there are also areas of flatter than usual land between the cratered regions. They are known as the intercrater plains.

RANGES AND RIDGES

The mountains and cliffs on Mercury tell us much about its past.

Four billion years ago, the bombardment of Mercury by lumps of rocks from space was at its height. One day, a particularly big lump slammed into the planet and struck it a devastating blow. It probably measured as much as 150 kilometres across and weighed millions of tonnes.

The rock crashed down on Mercury with unbelievable force and smashed to pieces. At the same time it gouged out a gigantic crater. Vast amounts of broken and molten rock blasted out of the crater shot kilometres above the surface, then rained down all around.

△ The circular ridges and mountains in the 1300-km wide Caloris Basin, formed when a gigantic meteorite hit Mercury.

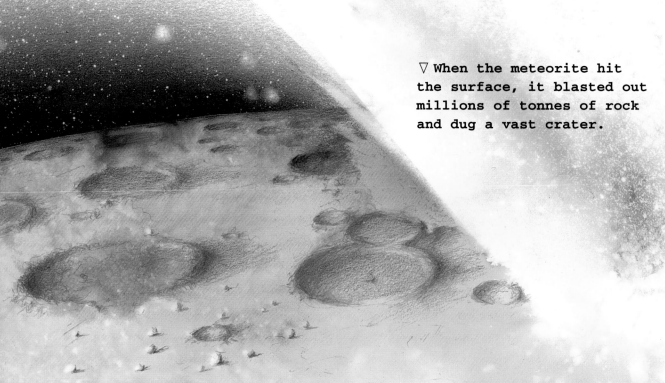

▽ When the meteorite hit the surface, it blasted out millions of tonnes of rock and dug a vast crater.

What a shock

The force of the impact was so great that it sent powerful shock waves through the surface rocks. The waves made the surface ripple, like the surface of a pond ripples when you throw in a stone.

But on Mercury, the ripples did not smooth out again after the shock waves had past. They formed rings of mountains, which we find on the planet today. The area within the rings of mountains is called the Caloris Basin, and it measures 1300 kilometres across. The rings of mountains, named Caloris Montes, are between about one and two kilometres high.

▽ **The force of the meteorite hitting Mercury sent shock waves through the surface and through the interior.**

The story of this incredible battering does not end here. The impact not only sent shock waves through the surface rocks. It also sent waves rippling right through the heart of the planet. And at the surface exactly opposite where the impact took place, all the shock waves came together.

Here they raised the ground level a kilometre or more and threw up little hills and ridges. Astronomers call this region weird or peculiar terrain. There is no other region like it anywhere else on Mercury.

Wrinkle ridges

Mercury's surface has also wrinkled up in many other places to form long ridges, or cliffs. They are known as scarps or rupes. They can be up to about three kilometres high, and may snake across the landscape for hundreds of kilometres. Similar scarps are found on the Earth. They formed when blocks of rock were forced upwards by movements in the crust.

surface shock waves

meteorite hits here

crust

interior shock waves

weird terrain

▷ **The interior shock waves disturbed the opposite side of Mercury, creating a strange or weird landscape.**

VENUS: THE SHROUDED PLANET

Thick clouds shroud Venus and prevent us seeing its surface.

△ The symbol for Venus.

Venus is the easiest planet to see. On many evenings we can see it shining in the west at sunset as the evening star. And early risers may see it in the east at dawn as a morning star.

Early astronomers thought that the morning and evening stars were different bodies. The Greeks called the morning star Phosphorus and the evening star Hesperus. But, because this planet looks so beautiful, it came to be called after Venus, the Roman goddess of beauty and love.

◁ A beautiful marble statue of the goddess of beauty. It is known as the Venus of Capua, the ancient Roman town where it was found.

Bright shiner

Venus is brighter than all the other stars and all the other planets. The only object brighter in the night sky is the Moon. There are two reasons why Venus shines so bright. One, it is the planet that comes closest to Earth – within 42 million kilometres. And two, Venus is covered with an atmosphere of swirling white clouds that reflect sunlight brilliantly.

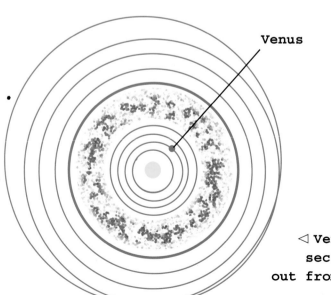

Venus

◁ Venus is the second planet out from the Sun.

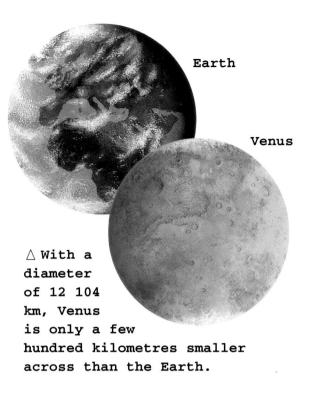

△ With a diameter of 12 104 km, Venus is only a few hundred kilometres smaller across than the Earth.

Slow spinner

Venus lies closer to the Sun than the Earth and does not take as long to circle it. Venus takes a little under 225 days to make the journey, compared with the Earth's 365 ¼. Unlike Mercury, Venus travels in nearly a perfect circle as it orbits the Sun.

If you lived on Venus, its slow spin would mean that you would only see the Sun rise every 117 Earth-days. And the Sun would rise in the west not in the east as it does on the Earth. This is because Venus spins round in space from east to west, in the opposite direction from all the other planets.

△ Venus has a surface shaped by volcanoes. The bright region in the middle of the picture is one of Venus's highland areas, Aphrodite Terra.

Hot stuff

Venus lies twice as far away from the Sun as Mercury, yet the planet is even hotter! Temperatures on Venus's surface may rise as high as 480°C, which is nearly ten times as hot as it ever gets on Earth. Why does Venus get so hot? It is because it has a very thick atmosphere that holds in the heat from the Sun.

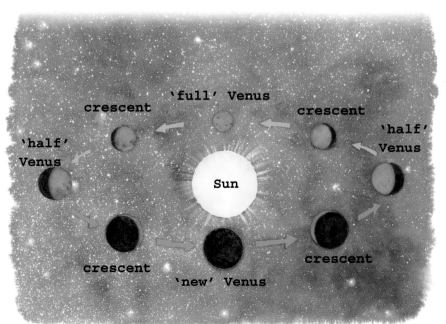

◁ Venus appears to change shape as it circles the Sun. This happens because we see different areas of its surface lit up by the Sun at different times. We call these changing shapes its phases. In turn, Venus goes from a crescent to full circle, then back to a crescent.

MAPPING VENUS

A varied landscape lies beneath Venus's clouds.

To the naked eye, Venus is a lovely object in the night sky. But it is a disappointing object in a telescope. All you can see of Venus is a shining white disc, faintly tinged with yellow. Thick clouds permanently hide the surface underneath.

Before the early 1970s, no one had any idea what the surface might be like. Then astronomers began using radio telescopes as radar stations to explore Venus' surface. Radar works by sending out pulses of radio waves and picking up echoes, or reflections, after the waves have been reflected back by an object. Unlike light waves, radio waves can pass through clouds.

The astronomers beamed waves through Venus's atmosphere and scanned across the surface. From the echoes they picked up, they were able to gain a general idea of the landscape. They identified plains regions, mountains and even faults, or breaks in the crust.

△ The space probe *Magellan*, which mapped the whole surface of Venus using radar beams.

◁ *Magellan* is launched from the space shuttle *Atlantis* on May 4, 1989. With its rocket booster it weighs more than 3 tonnes.

Space invaders

By the late 1970s space probes from Russia and the United States had begun to explore Venus. The probe *Pioneer Venus Orbiter* was orbiting the planet and scanning its surface with radar. It was beginning to reveal the surface of Venus clearly for the first time. The probe stopped working in 1992. By then another probe, named *Magellan*, was circling the planet. With its better radar, *Magellan* was able to picture the planet in far greater detail.

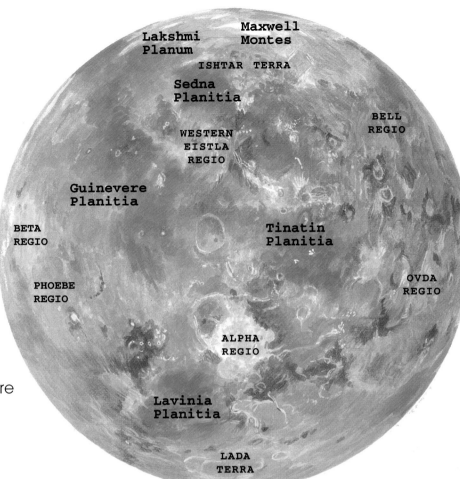

Maxwell Montes
Lakshmi Planum
ISHTAR TERRA
Sedna Planitia
BELL REGIO
WESTERN EISTLA REGIO
Guinevere Planitia
Tinatin Planitia
BETA REGIO
OVDA REGIO
PHOEBE REGIO
ALPHA REGIO
Lavinia Planitia
LADA TERRA

Atalanta Planitia
Vinmara Planitia
Niobe Planitia
ULFRUN REGIO
Rusalka Planitia
ALTA REGIO
THETIS REGIO
Diana Chasma
APHRODITE TERRA
Artemis Chasma

Detailed maps

By the time *Magellan*'s mission ended, in October 1994, it had mapped the whole planet in detail and showed that Venus is one of the most fascinating bodies in the whole Solar System.

◁ △ These maps are based on the images sent back by *Magellan*. Some large craters are named. Planitia are plains; terra is a continent; mons (plural montes) is a mountain; regio is an upland region; chasma is a rift, or deep valley.

THE DEADLY ATMOSPHERE

Acid clouds float in Venus's suffocating atmosphere.

Venus is often called Earth's twin or sister planet. But it is only similar to Earth in size. In every other way, Venus is quite different. In particular, Venus's atmosphere is very different from Earth's because it is very thick. This means that it contains more gas than the Earth's atmosphere.

The main gas in Venus's atmosphere is carbon dioxide. There is also a little nitrogen and argon, and traces of other gases too. Carbon dioxide is much denser, or heavier, than air, and this makes the atmospheric pressure on Venus more than 90 times the atmospheric pressure of the Earth. An astronaut venturing out on Venus's surface in an ordinary spacesuit would be crushed to death.

▽ **The *Pioneer* Venus probe showed curious Y-formations among the clouds on Venus.**

◁ **The clouds of *Venus* show up in this picture taken by the *Galileo* probe in infrared light.**

Venus's greenhouse

The astronaut would be baked at the same time because the temperature on Venus can reach 480°C. This high temperature is also caused by the presence of the carbon dioxide in the atmosphere. The gas turns the planet into a kind of greenhouse, which holds in the heat just like a garden greenhouse.

A similar greenhouse effect is taking place on Earth because the amount of carbon dioxide in the atmosphere is increasing. It is slowly causing the climate to warm up.

▽ **This picture taken by *Mariner 10* was the first to show how the clouds circulate in Venus's atmosphere.**

Clouds and Winds

The clouds on Venus form much higher up than clouds do on the Earth. On the Earth, clouds can start forming less than 1000 metres above the ground. But on Venus, the clouds do not begin until nearly 50 kilometres above the surface. And they reach up to more than 70 kilometres.

At the top of the clouds, the wind blows at hurricane force, reaching speeds of 350 kilometres an hour. The winds whip the clouds round the planet in just four days, travelling from east to west. Yet the planet takes eight months to spin round once! The wind gradually drops lower down, and near the surface it is little more than a gentle breeze.

VENUSIAN LANDSCAPES

Most of Venus is made up of rolling plains formed from great lava flows.

Underneath the clouds, Venus has two main kinds of land regions – plains and highlands. Plains cover at least 85 per cent of the planet's surface, with highland areas making up only 15 per cent.

The plains regions (planitia) have been formed by flows of lava that have poured out from the many volcanoes that dot the surface. These volcanoes have erupted time and time again, each time sending out new lava flows to cover the old. Repeated lava flows can be seen in the images sent back by the *Magellan* probe.

The shields of Venus

Most of the volcanoes on Venus seem to be similar to a type found on Earth we call a shield. Shield volcanoes give off very runny lava that can travel a long way very quickly. This results in a broad, low volcano rather than a tall, cone-shaped one. There are literally hundreds of shield volcanoes on Venus, some of them measuring more than 400 kilometres across. Only on Mars have bigger volcanoes been found.

△ **One of Venus's big volcanoes, Sif Mons. It is about 2 km high and 300 km across.**

▷ **Vast lava flows surround a volcano that has recently erupted.**

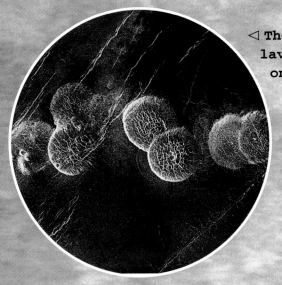

◁ These flat, circular lava 'pancakes' are only found on Venus.

Deep rifts

Aphrodite Terra is nearly three times as big as Ishtar Terra and covers an area about the size of Africa. It is dotted with huge volcanoes, such as the 9-kilometre high Maat Mons.

In the eastern part of Aphrodite Terra long, deep valleys are found. They are called rift valleys, or chasmata (singular chasma). They are great gashes in the surface caused by massive movements of Venus' crust. They include the horseshoe-shaped Artemis Chasma and the nearby Diana Chasma, which is 4 kilometres deep.

Chasmata are also found elsewhere on Venus. Devana Chasma is one of the biggest. It is 90 kilometres wide and runs for 2500 kilometres between Beta Regio and Phoebe Regio.

Continental features

The two main highland regions on Venus are often called the planet's continents. There is one in the north, called Ishtar Terra, and one near the equator, called Aphrodite Terra. There are other smaller highland areas elsewhere, such as Beta Regio and Phoebe Regio.

Ishtar Terra is the smaller of the two continents, measuring about 3000 kilometres across. In area, it is about the same size as Australia. Two main landforms dominate Ishtar. One is a huge plateau called Lakshmi Planum. The other is a mountain range, Maxwell Montes, the highest on the planet. Some peaks rise to more than 12 kilometres high.

▷ This false-colour picture of Venus shows the mouth of a volcano (bottom right). The different colours around it are lava flows from a number of eruptions.

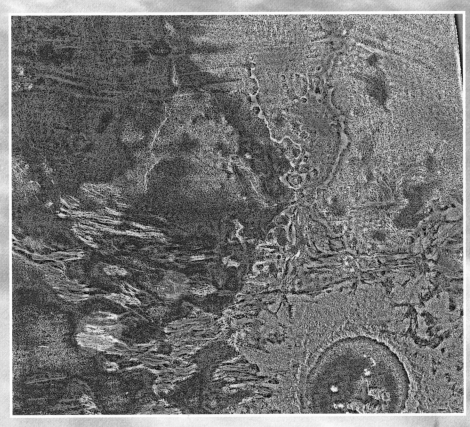

FANTASTIC FEATURES

Volcanic activity, crust movements and meteorites have scarred Venus's surface.

The lava flows we see around many of Venus's volcanoes are like the lava flows we see around volcanoes on the Earth. But some volcanic activity on Venus creates other features that are never found on the Earth, or anywhere else in the Solar System.

▽ **Networks of fine fractures looking like cobwebs are found in many places.**

Examples of these features are the pancake domes. These are flat-topped 'bubbles' of lava, which look like pancakes. Geologists think that they formed when viscous (thick) lava forced its way out of vents, or holes, in flat ground. Because the ground was flat, the lava formed circles.

Crowns and tiles

Coronae are other features found only on Venus. The word corona means crown, and a Venus corona is a crown-like, circular region, a bit like a crater. Around a raised centre is a depression, like a moat around a castle. Around the 'moat' the ground rises to a rim, which is marked by a series of cracks, or fractures.

▽ **The magnificent volcano Maat Mons, named after an Egyptian goddess. Its peak is more than 6 km high.**

Geologists think that coronae are regions where molten rock from underground has pushed up the surface, and then drained away, causing the surface to collapse. The circular fractures show where the rocks have cracked when the ground moved.

Other cracks, faults and ridges are found all over the planet. The most unusual fault system, again found only on Venus, is called a *tessera*. The word means tiles, and a Venus *tessera* is a criss-cross network of fractures, that looks like a pattern of tiles on a kitchen floor.

Venus's craters

Like all the near planets, Venus has been bombarded by meteorites throughout its history. They have dug out craters in the surface. But Venus has few craters compared with Mercury, for example. There are two reasons for this. One, Venus has a thick atmosphere, which helps protect it from meteorites. Two, lava flowing from its many volcanoes has hidden most old craters.

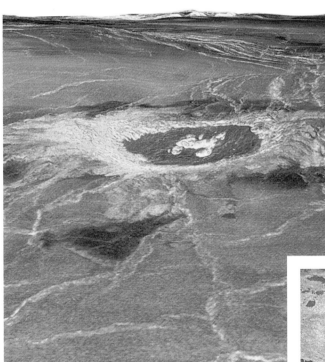

△ **One of the few large meteorite craters found on Venus. It is nearly 50 km across.**

▷ **An example of the peculiar fractured landscape known as *tessera*.**

MARS: THE RED PLANET

Mars is similar to Earth in some ways, but very different in others.

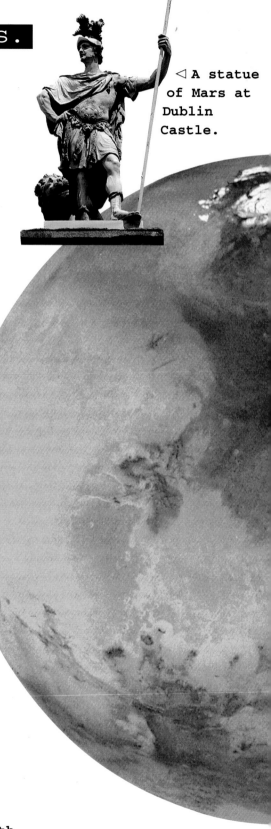

◁ A statue of Mars at Dublin Castle.

When it is bright, Mars is a very easy planet to recognize because it shines with a noticeable orange-red colour. To ancient peoples, its reddish colour suggested fire and blood. That is why it came to be called Mars, after the Roman god of war.

The brightness of Mars varies a lot because its distance from the Earth varies widely. It is brightest at opposition. This is when it is opposite the Sun in the heavens and closest to the Earth. Then it glows like a fiery beacon. Oppositions take place about every two years. There is a particularly favourable opposition every 15 years or so, and on these occasions Mars can come within 56 million kilometres of the Earth – nearer than any planet except Venus.

△ The symbol for Mars. It includes a representation of the god of war's spear.

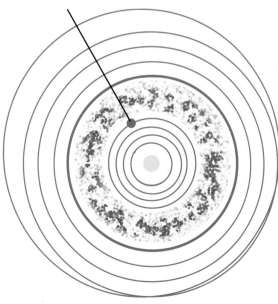

Mars

▷ A picture of Mars taken by the Hubble Space Telescope, showing the northern ice cap.

◁ Mars is the fourth planet out from the Sun.

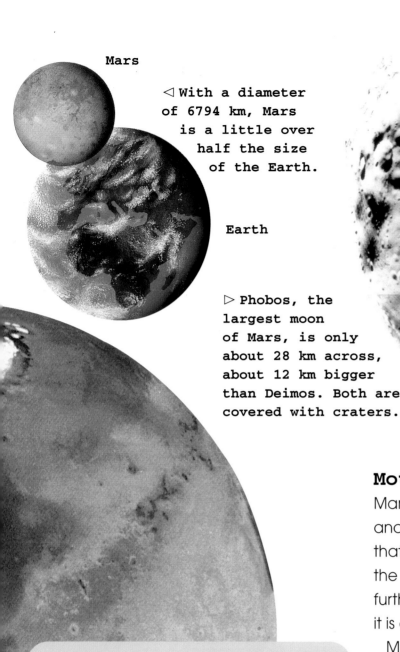

Mars

◁ With a diameter of 6794 km, Mars is a little over half the size of the Earth.

Earth

▷ Phobos, the largest moon of Mars, is only about 28 km across, about 12 km bigger than Deimos. Both are covered with craters.

Martian moons

Mars has two tiny satellites, or moons. They are not like our own Moon, and are really nothing more than two big shapeless lumps of rock. They are named Phobos (Fear) and Deimos (Terror), after the two horses that pulled the chariot of the war god Mars. Astronomers think that these two moons could be asteroids that Mars captured from the nearby asteroid belt.

Motions of Mars

Mars is the fourth planet out from the Sun and lies outside the Earth's orbit. This means that it takes longer to travel once around the Sun – 687 Earth-days. Because Mars is further away from the Sun than the Earth, it is a much colder planet.

Mars spins on its axis once every 24 hours 37 minutes. This means that a day on Mars is only about half an hour longer than a day on Earth. The planets are also similar in another respect. The axis of Mars is tilted in space, just like the Earth's axis.

Martian seasons

On the Earth, the tilted axis causes the seasons, or the gradual and regular changes in the weather over the year. The tilted axis of Mars brings seasons to that planet too – it has a spring, summer, autumn and winter. And they bring about regular changes in the Martian weather.

mapping mars

MAPPING MARS

Mars's surface abounds with fascinating features.

Through a telescope Mars is more interesting than the other near planets. We can see that it has ice caps at the north and south poles. And we can see dark markings elsewhere, which change all the time. But Mars is too far away for us to see the surface clearly.

The space probe *Mariner 4* provided the first close-up views of Mars in 1965. The few pictures it sent back showed a cratered landscape. Perhaps Mars was going to turn out to be like the Moon. But later *Mariner* probes revealed that craters cover only part of Mars's surface.

When two *Viking* probes began scanning the planet in detail in 1976, a whole variety of surface features came into view – vast sandy plains, rugged highlands, towering volcanoes, deep canyons, and winding channels that look like rivers on Earth.

▷ **In 1976, two *Viking* orbiters dropped capsules to land on the surface.**

△ **Background: 1997, the *Pathfinder* probe parachuted down to Mars inside an airbag, which cushioned its landing.**

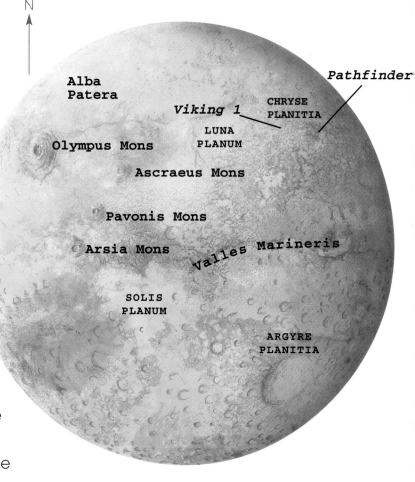

▽▷ These maps are based on the images sent back by the *Viking* orbiters. Some large craters are named. Planitia are plains; mons (plural montes) is a mountain; vallis is a valley; mensa is a flat-topped plateau; fossa (plural fossae) is a shallow depression. The landing sites for the *Viking* landers and *Pathfinder* are marked.

Happy landings

The *Viking* probes that scanned Mars were orbiters, which circled the planet in orbit. They also dropped landing probes that touched down on the surface and took close-up photographs of rocks and soil. These landers also checked on the local weather, and carried equipment to analyse, or find out the make-up of, the soil.

The landers had a digging arm that scooped soil into a miniature laboratory. This tested the soil using an X-ray instrument. Other experiments tested the soil for organic matter, to see if there was any life on Mars.

Another probe, called *Pathfinder*, landed on the planet in July 1997. It carried a tiny rover called *Sojourner*, which roamed about the planet for several weeks. This had a camera to take close-up pictures of rocks and also an X-ray instrument to analyse them. By the time *Pathfinder* had sent its final signals, in September 1997, another probe, called *Global Surveyor*, was swinging into orbit around Mars. Soon it began taking the finest pictures ever of the once mysterious Red Planet.

◁ These *Pathfinder* pictures show the Sun rising over the horizon (top left) and various cloud formations in the dusty Martian sky.

THE MARTIAN WEATHER

The weather on Mars changes day by day and season by season.

A typical weather forecast on Mars might not be so different from a typical weather forecast on Earth. 'Early morning frost and mist will soon disappear. Strong winds will clear the clouds from highland areas, but there may be storms later.'

Mists, clouds, winds, storms, frosts and ice all play their part in deciding the weather on Mars, just as they do on the Earth. This might seem surprising considering how different the atmospheres are on the two planets. Mars's atmosphere is much thinner, which means that the planet has only a small amount of gas around it compared with the Earth.

▽ This *Viking* image shows a crescent Mars and some of its features. At left, clouds gather around Olympus Mons volcano. In the centre is the great gash of Mariner Valley. At right is a frost-filled Argyre Basin.

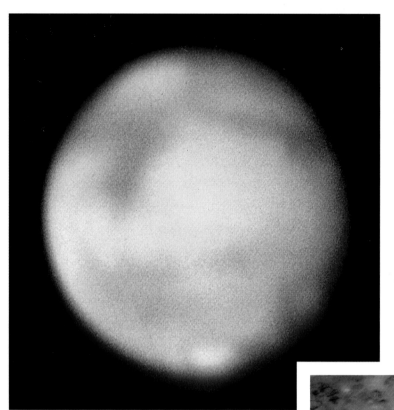

◁ A telescope view of Mars, showing the two polar ice caps. The caps come and go as the seasons change.

Martian winds

Winds blow for much of the time on Mars, even though its atmosphere is so thin. At times they can gust up to 120 kilometres an hour, which would be a strong gale on Earth. They pick up fine dust from the surface and whirl it into the air, creating huge dust storms. Sometimes these storms rage all round the planet for weeks at a time.

△ The arrow points to a dust storm developing in a desert-like region.

Changing temperatures

Overall, Mars is much colder than the Earth because it is much further away from the Sun. But the temperature varies widely, depending on the location, the time of day and the season. The highest temperatures are reached near the equator at noon in summer. There, they can climb as high as 25°C, which would be a comfortable temperature on Earth. But for most of the time over most of the planet, midday temperatures stay below freezing.

And everywhere, as soon as the Sun goes down, the temperature falls sharply. This is because the atmosphere is too thin to hold in much heat. During the night the temperature can fall to –100°C or below, far colder than it ever gets on Earth. Temperatures are even lower at the poles, where they can fall as low as –150°C in winter.

Mists and clouds

Mists often appear in the valleys and canyons of Mars after a frosty night. The Sun evaporates the icy frost into water vapour. This turns into a mist of tiny water droplets, which disappears when the atmosphere warms up. The higher-level clouds that are sometimes seen around the mountains on Mars are made up of water-ice crystals, just like high cirrus clouds on Earth.

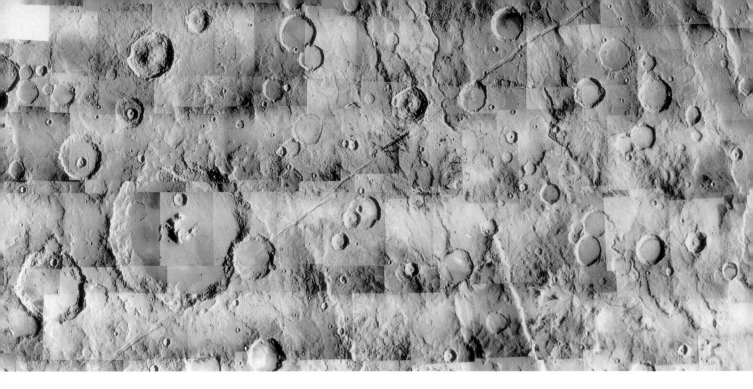

MARTIAN LANDSCAPES

Mars boasts the biggest volcano and the longest valley in the Solar System.

Mars is a planet of great contrasts. There are low rolling hills, towering volcanoes, vast smooth plains, rugged regions peppered with hundreds of craters, shallow channels snaking through the plains, and canyons many kilometres deep.

There is also an overall contrast between the northern and southern parts of Mars. The southern part of the planet is heavily cratered, while the northern part consists mostly of plains. The cratered area in the south is probably the oldest region of Mars, which formed billions of years ago when meteorites bombarded the planet.

Two particularly large meteorites or asteroids gouged out huge basins. The biggest is Hellas, which measures more than 1500 kilometres across. It is about twice as big as the other, Argyre.

There are relatively few craters in the northern part of Mars. And in general they are smaller and younger than the craters in the south. Much of the surface consists of vast plains (planitia), rather like some of the desert areas on Earth. The plains have almost certainly been formed from the great lava flows from ancient volcanoes.

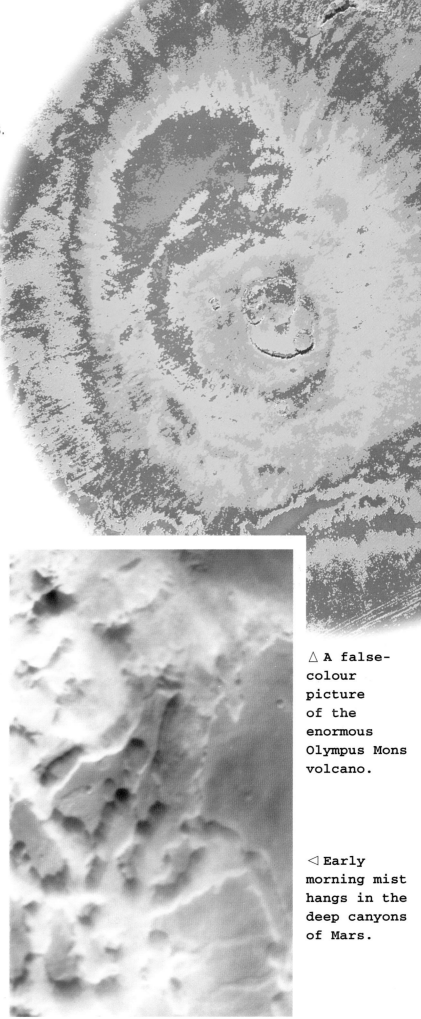

Mount Olympus

There are plenty of volcanoes on Mars. Some of the biggest are found in a region called Tharsis, close to the planet's equator. The whole region forms a great bulge on the surface of Mars, and includes three large volcanoes up to 20 kilometres high.

But the Tharsis trio are dwarfed by an even larger volcano, which rises to a height of 25 kilometres. Called Olympus Mons (Mount Olympus), it is the biggest volcano in the Solar System. At its base, it measures 600 kilometres across, which makes it as big as England and Wales.

Mariner Valley

Mars has another record-breaking natural feature, which has been called the Martian Grand Canyon. It begins not far away from the Tharsis Ridge and runs close to Mars's equator for nearly 5000 kilometres. It was named Valles Marineris (Mariner Valley) after the *Mariner 9* space probe that first spotted it.

Mariner Valley is much longer than the Grand Canyon in the USA, and is also deeper and wider. It was also formed in quite a different way. The American Grand Canyon was formed by erosion, as the Colorado River cut through the surface rocks. Mariner Valley, on the other hand, is a geological fault – a crack caused by movements in Mars's crust.

△ **A false-colour picture of the enormous Olympus Mons volcano.**

◁ **Early morning mist hangs in the deep canyons of Mars.**

37

ON THE PLAINS

Rocks are scattered everywhere on the rust-red Martian plains.

Planitia, or plains, cover much of the Martian surface. Space probes have landed on two of the plains and have revealed that they look like stony deserts here on Earth. The *Viking 1* lander (see page 33) provided us with our first close-up views of the Martian surface. It came down on a plain known as Chryse Planitia, or the Plain of Gold. Its cameras revealed a gently rolling landscape strewn with small rocks, with one or two large ones dotted about here and there.

All the rocks were embedded in a kind of sandy soil. In places the soil had gathered into drifts like miniature sand dunes in the desert. The soil and the rocks were a rich orange-red.

▽ **An all-round, panoramic view of *Pathfinder's* landing site, made up of four separate pictures.**

Same scenes

Viking 2 landed on another plain, Utopia Planitia, hundreds of kilometres away from *Viking 1*. But when its cameras peered at the surface, they showed it to be almost identical to Chryse. Rock embedded in sandy soil littered the landscape as far as the horizon in all directions.

The third probe to land on Mars, *Pathfinder*, also set down on Chryse, about 800 kilometres from *Viking 1*. It landed in what is thought to be an ancient river bed at the mouth of Ares Vallis, or Mars Valley.

The pictures taken by *Pathfinder's* cameras were similar to those taken by the *Vikings*. But they showed a greater variety of rock types and more soil. This is what would be expected in a river bed.

△ The Martian surface, as viewed by the *Viking 1* lander on Chryse.

▷ The rocks scattered around *Pathfinder* were pitted and covered with soil.

Regarding rocks

The rocks that litter the Martian plains were almost certainly not formed where we find them today. So where did they come from? There are three main processes that could have been at work. Volcanoes might have flung them out when they erupted. They might have been hurled out from craters dug by meteorites. Or ancient rivers might have transported them.

Most Martian rocks seem to be volcanic, and have proved to be surprisingly similar to volcanic rocks we find on Earth, such as basalt and andesite. But, some of the rocks look as if they could be conglomerate. This is a typical sedimentary rock made up of rounded pebbles cemented together. This type could only have been formed with the help of flowing water.

LIFE ON MARS

Primitive life forms may once have existed on Mars.

As far as we know, the Earth is the only planet in the Solar System on which there are living things. Earth's neighbours in space seem to be either too hot, too cold, or too dry for life as we know it to exist.

Until the1970s, no one really knew what conditions were like on other planets, and many people thought that Mars might be warm enough to support some form of life.

Martian canals

In 1877, an Italian astronomer named Giovanni Schiaparelli, said he had spotted 'canali' on Mars. In Italian, the word canali means channels, but many people thought it meant canals, or artificial waterways built by intelligent beings. So the idea grew up of a race of Martian people who had built canals to carry water from the ice caps at the poles to warmer regions so that they could grow crops.

Even though more powerful telescopes showed few signs of canals, the idea of Martians remained popular. Even today, some people still believe that Martians visit Earth in flying saucers, or UFOs (unidentified flying objects).

Background: One of the sketches of 'canals' on Mars drawn by the Amercan astronomer Percival Lowell. He was convinced that there were Martians.

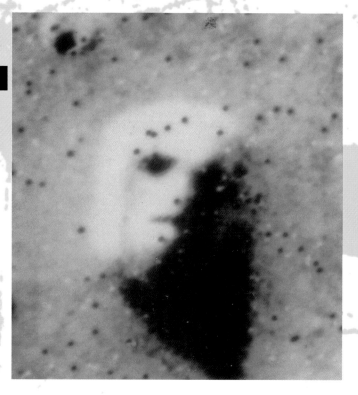

△ **The face of a Martian?** *Viking* **pictures showed what looks like a face carved in the rocks. But it is only a natural feature.**

Martian microfossils

In 1996, NASA scientists claimed that they had discovered the first signs of life on Mars. They said that they had found microfossils in a meteorite (ALH84001) that came from Mars. The tiny fossils, visible only under a microscope, were the remains, they said, of bacteria. But many scientists think that the 'fossils' are peculiar mineral shapes and nothing to do with past life.

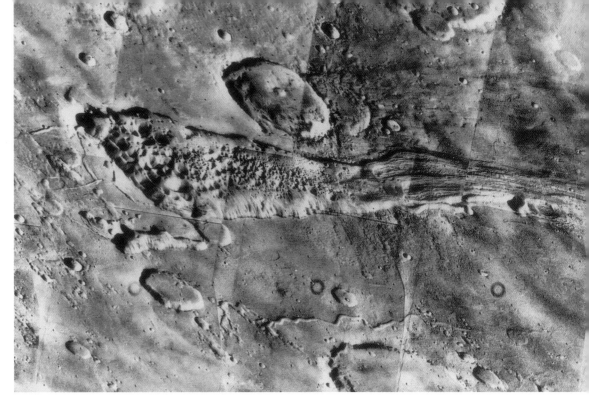

▷ **This large channel on Mars was probably made by water flooding the surface millions of years ago. There would have been plenty of water available for living things.**

Barren wasteland

But since then, space probes have shown that Mars is a cold, dry and barren wasteland. There are no canals, no vegetation, and definitely no intelligent Martians. The *Viking* landers also found no traces of living organisms.

But, pictures taken by the *Viking* and later probes have shown that water once flowed on Mars. This means that conditions were wetter and warmer in the past.

Maybe conditions were then suitable for life. And maybe some kind of life did develop, only to die out when conditions became harsh like they are today. Few planetary scientists would really be surprised if they found fossils in the samples of Martian soil that probes will return to Earth in a few years time.

▽ **Part of the north polar ice cap of Mars in summer, when much of it has melted.**

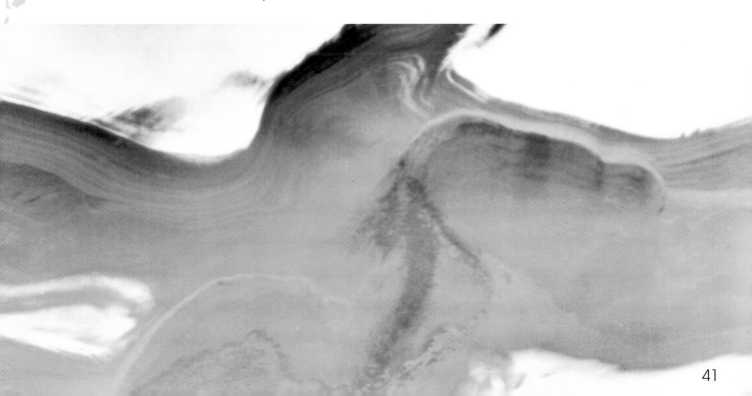

TIME LINE

300s BC

The Greek philosopher Aristotle is first to observe an occultation (covering up) of Mars by the Moon.

265 BC

The first observation of Mercury is made, according to the last great Greek astronomer Ptolemy of Alexandria, who lived around AD 150.

AD 60

The Roman historian Pliny records around this time that Venus sometimes casts shadows.

1610

Galileo becomes the first person to see the phases of Venus, when he pioneered telescopic observation of the heavens.

1600s

The Dutch astronomer Johannes Hevelius becomes the first person to spot the phases of Mercury during telescopic observations early in the century.

1631

The French astronomer Pierre Gassendi is first to observe a transit of Mercury on November 7, as predicted by the German astronomer Johannes Kepler. Only 48 transits have occurred since.

1639

A transit of Venus is observed on November 24 by English amateur astronomers J Horrocks and W Crabtree. Only four transits have been observed since.

1659

The Dutch astronomer Christiaan Huygens is first to record a marking on Mars, which becomes known as Syrtis Major.

1727

In his novel *Gulliver's Voyage to Laputa*, Jonathan Swift writes that Mars has two satellites.

1877

The real satellites of Mars are discovered in August by the American astronomer Asaph Hall. The Italian astronomer Giovanni Shiaparelli reports seeing 'canals' on Mars.

1880s

Schiaparelli studies Mercury between 1881 and 1889 and produces the first map of Mercury, showing patterns of dark markings.

1894

American astronomer Percival Lowell sets up his Flagstaff Observatory to observe Mars and its 'canals'.

1896

Percival Lowell reports seeing 'canals' on Mercury. No-one else does.

1934

The Greek astronomer Eugenios Antoniadi publishes the best map of Mercury before the Space Age.

1954

The American astronomer Frank Whipple suggests that Venus is covered with a great water ocean.

1961

The first radar contact is made with Venus by the Lincoln Laboratory in the United States.

1962

Mariner 2 is launched in August, and in December becomes the first successful space probe to report back on another planet – Venus. It reports that the planet is very hot.

1965

The first close-up images of Mars are sent back by *Mariner 4* over a record communications distance of 220 million km.

1967

Russia's *Venera 4* parachutes into Venus's atmosphere and reports on conditions there.

1971

Mariner 9 discovers Mars's 'Grand Canyon', later named Mariner Valley in its honour, after becoming the first probe to go in orbit around another planet.

1974

Venus's cloud formations are photographed by *Mariner 10* in February. The probe goes on to take the first close-up photographs of Mercury in March and September. They are still the only ones available.

1975

The first pictures of Venus's surface are sent back by Russia's *Venera 9* probe, which lands on the planet in October.

1976

Two *Viking orbiters* go into orbit around Mars, and drop landers down to the surface in July and September.

1978

Large parts of Venus's surface are revealed for the first time by *Pioneer-Venus 1*, which scans the planet from orbit, using radar.

1990

Venus is mapped in detail by the *Magellan* probe from orbit, using radar.

1996

NASA scientists report that they have discovered microscopic fossils in a meteorite from Mars (ALH84001). Their discovery is later hotly disputed.

1997

Mars *Pathfinder* lands on Mars in July. It carries a wheeled rover, *Sojourner*, which analyses nearby rocks.

2004

There will be a transit of Venus on June 8, the first since 1882.

Mercury data

Diameter at equator: 4878 km

Volume: 0.1 Earth's volume

Mass: 0.1 Earth's mass

Density: 5.4 times density of water

Gravity at surface: 0.4 Earth's gravity

Distance from Sun average: 57 900 000 km

furthest: 69 700 000 km

closest: 45 900 000 km

Spins on axis in: 58.7 days

Circles Sun in: 88 days

Speed in orbit: 172 000 km/h

Temperature: -180° to 450°C

Moons: 0

Mercury notes

TWO FOR THE PRICE OF ONE

The only space probe that has visited Mercury, *Mariner 10,* flew past Venus first. It used the gravity of Venus to speed it up and fling it in the right direction. This was the first time that this technique, called gravity-assist, was carried out.

ONLY A WHIFF

Astronomers tell us that they have detected a slight atmosphere on Mercury. But they say that the pressure of the atmosphere is only one million millionth of the pressure of the atmosphere on Earth. They reckon that this atmosphere is made up of particles of helium gas given off by the Sun in the solar wind.

WHERE IS VULCAN?

Years ago astronomers thought that there might be another planet orbiting closer to the Sun than Mercury, and they called it Vulcan. We now know that there isn't.

MERCURY IN TRANSIT

Very occasionally Mercury can be seen crossing the surface of the Sun, an event called a transit. Fourteen transits occurred in the 20th century, the last in November 1999. Transits always occur either in November or in May. May transits can last for up to 9 hours.

ANY OLD IRON

Mercury contains in its core more iron for its size than any other planet. Its core is bigger than our Moon.

EASY TO SWALLOW

When the Sun begins to die in 5000 million years time, it will swell up to become a red giant. As it expands, it will swallow Mercury. The planet will first melt then turn to gas in the searing hot interior of the expanding Sun.

Venus data

Diameter at equator: 12 104 km

Volume: 0.9 Earth's volume

Mass: 0.8 Earth's mass

Density: 5.2 times density of water

Gravity at surface: 0.9 Earth's gravity

Distance from Sun average: 108 000 000 km

      furthest: 109 000 000 km

      closest: 107 000 000 km

Spins on axis in: 243 days

Circles Sun in: 224.7 days

Speed in orbit: 126 000 km/h

Temperature: 450°C

Moons: 0

Venus notes

CASTING SHADOWS

Venus sometimes shines so bright that it can be seen during the daytime. And sometimes at night it will cast shadows. The Roman writer Pliny wrote about this 2000 years ago.

VENUS IN TRANSIT

Like Mercury, Venus can be seen crossing the face of the Sun. But these transits are even rarer than those of Mercury. There were no transits at all in the 20th century. The last was in 1882, and the next will not be until 2004.

WHAT A WAY TO GO

Venus is one of the most unpleasant places in the Solar System. If you were to travel there, you would be at the same time crushed to death by the terrific pressure in the atmosphere, baked to death in the furnace-like temperatures, and burned to death by the sulphuric acid mists.

TERRAFORMING VENUS

Scientists have come up with an ingenious way of turning Venus into another Earth, a process called terraforming. First they would send tiny organisms called algae into the atmosphere. These would feed on the carbon dioxide and give off oxygen. As the carbon dioxide is used up, the greenhouse effect would cease, allowing the atmosphere to cool. Eventually it would cool so much that the moisture in the atmosphere would fall to the surface as rain, bringing about further cooling. After thousands of years – it is a long term project! – the planet would cool enough to allow life as we know it to gain a hold.

Mars data

Diameter at equator: 6794 km

Volume: 0.2 Earth's volume

Mass: 0.1 Earth's mass

Density: 3.9 times density of water

Gravity at surface: 0.4 Earth's gravity

Distance from Sun average: 228 000 000 km

      furthest: 249 000 000 km

      closest: 207 000 000 km

Spins on axis in: 24 hours 36 minutes

Circles Sun in: 687 days

Speed in orbit: 86,000 km/h

Temperature: -120° to 25°C

Moons: 2

Mars notes

MARTIAN METEORITES

Most meteorites are bits of rock that have been wandering around the Solar System since it began. But some are thought to come from Mars, blasted into space from the surface when it was struck by a huge meteorite.

WAR OF THE WORLDS

In 1938, an American actor/producer named Orson Welles created panic when he broadcast a play based on H G Wells's novel *War of the Worlds*, which told of Martians waging war on the Earth. Many listeners didn't realize it was a play and thought Earth was really under attack.

GLOSSARY

arachnoid
A strange, spider-like feature on the surface of Venus.

asteroids
Lumps of rock that orbit the Sun in a band between the orbits of Mars and Jupiter.

astronomy
The scientific study of the heavenly bodies.

atmosphere
A layer of gases around a planet or a star.

atmospheric pressure
The weight of the atmosphere pressing down.

axis
An imaginary line around which a body spins.

canals
Channels astronomers once reported seeing on the surface of Mars.

chasma
A kind of deep valley.

chemical elements
The basic chemicals found in all kinds of substances; the building blocks of matter.

comet
An icy ball of matter that starts to glow when it nears the Sun.

constellation
A group of bright stars that appear in the same part of the sky.

core
The centre part of a planet or moon.

corona
A circular feature found on Venus.

crater
A pit dug out of the surface of a planet or moon, caused by a meteorite.

crust
The hard outer layer of a rocky planet.

day
The time it takes the Earth to spin round once on its axis in space. A planet's 'day' is the time the planet takes to spin round once on its axis.

equator
An imaginary line around a planet or a moon, midway between the north and south poles.

erosion
The gradual wearing away of the landscape, for example by the weather or flowing water.

evening star
Usually the planet Venus appearing in the western sky after sunset. Mercury can be an evening star as well.

galaxy
A great star 'island' in space, containing many billions of stars.

gravity
The force that attracts one lump of matter to another.

greenhouse effect
What happens when a planet's atmosphere acts like a greenhouse and traps the Sun's heat.

mantle
The layer of rock beneath the surface crust of a rocky planet.

maria
Flat plains on the Moon, commonly called seas.

meteorite
A piece of rock or metal from outer space that falls to the ground.

moon
A natural satellite of a planet.

morning star
Usually the planet Venus shining in the eastern sky just before sunrise. Mercury can be a morning star as well.

orbit
The path in space a heavenly body follows when it circles another.

phases
The different shapes of Mercury and Venus (and the Moon) when we see more or less of them lit up by the Sun.

planet
A large body that circles in space around the Sun.

planitia
Flat plains regions on a planet.

polar caps
Layers of ice seen at the north and south poles of Mars.

radar
A method astronomers use to picture distant planets by bouncing radio waves off them and recording the 'echoes'.

satellite
A body or an object that circles around another body in space. The Moon is a natural satellite of the Earth. The Earth also has many artificial satellites.

season
A period of the year when temperatures and the general weather are much the same as in previous years. Mars has seasons as well as the Earth.

solar
To do with the Sun.

Solar System
The Sun's family, which includes the planets and their moons, asteroids and comets.

space probe
A spacecraft that escapes the Earth's gravity and travels to other planets and their moons.

star
A huge ball of hot gases, which gives off energy as light, heat and other radiation.

terrestrial
Like the Earth.

transit
When Mercury or Venus can be seen moving across the face of the Sun.

Universe
Everything that exists – space, stars, planets and all the other heavenly bodies.

volcano
A place where molten rock forces its way to the surface of a planet or moon.

year
The time the Earth takes to circle once in space round the Sun. A planet's 'year' is the time it takes the planet to circle round the Sun.

zodiac
An imaginary band in the heavens in which the Sun and the planets are always found.

INDEX